Microbiology Experiments You Can Do at Home

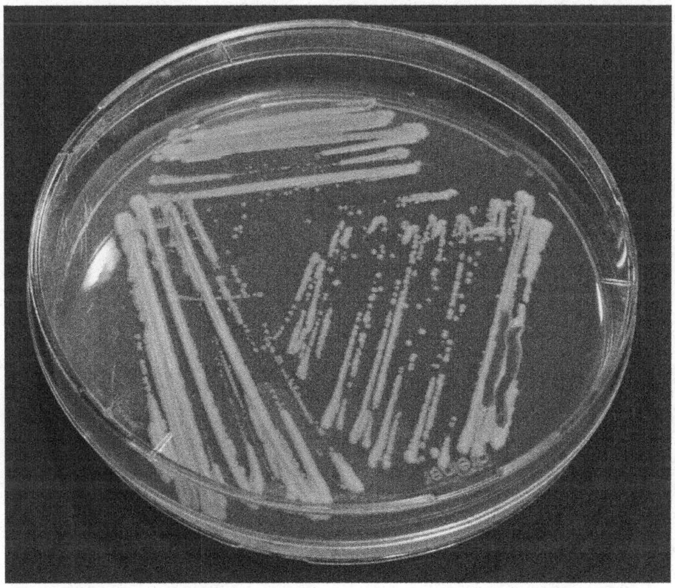

Dr. Caroline Crocker

ISBN 978-1-957970-14-1 (ebook)
ISBN 978-1-957970-13-4 (paperback)

Publisher: Rambling Ruminations, Fairfax, VA, USA

Illustrations are primarily from the author's photographs but also include some public-domain photographs.

Author blog: https://ramblingruminations.com

YouTube playlist: https://tinyurl.com/CrockerMicro

Table of Contents

Instructions for Use of This Manual

This manual was originally written as a supplement to the online lab exercises that formed part of a college-level Introduction to Microbiology course taught during COVID lockdowns. As such, it is suitable for use with any introductory college-level biology course where students must work at home. However, with adult supervision, home-schooled high school and even junior high school students can easily complete the experiments and learn perhaps more than they want to know about the bugs and germs around them.

Please note that every effort has been made to allow students to experience microbiology safely. At no time do the students open their agar plates, and neither should they. Such manipulations must only be done in a microbiology lab.

To complete the experiments described in this manual, students will need to purchase either a sleeve of pre-poured agar suitable for growing bacteria (Nutrient agar or LB agar) or a sleeve of 10 plates of pre-poured agar suitable for growing fungi (Sabouraud, Potato dextrose or Malt Extract). They will also need to purchase a sleeve of MRS agar plates for growing Lactobacilli. All of these are available on Amazon. These plates should be refrigerated until used. Students will have extra plates for growing bacteria. Those should be kept in case of need. Mistakes happen. Students will also need a Sharpie, tape, bottled water, a black plastic bag, a metal tin that can house plates, and various other household items, as listed in the Methods sections.

Instructors/homeschoolers will notice that the explanation given at the beginning of the lab is extremely brief. It is expected that they will supplement this when they are going over the instructions to prepare students to complete the exercise. In addition, at the end of each lab, there is an Analysis section where students are given instructions on how to turn in a lab report. Again, instructors are encouraged to differ from this if they believe it will improve the students' learning experience.

Finally, this lab manual is a work in progress and is in the process of being supplemented and revised for optimal use. Therefore, any suggestions for improvement will be gratefully received.

Disclaimer

These experiments have been designed with safety in mind. However, Dr. I. Caroline Crocker (the Author) assumes no responsibility for possible illness, disability, or even death resulting from performing or altering the experiments herein.

To the maximum extent permitted by applicable law, the Author shall not be liable for any indirect, incidental, special, consequential or punitive damages, or any loss of profits or revenues, whether incurred directly or indirectly, or any loss of health, resulting from (1) your use of Microbiology Experiments You Can Do at Home (this Manual), (2) your performing of experiments contained herein, (3) or the performing of the experiments by any others.

The Author assumes no responsibility for errors, inaccuracies, omissions, or any other inconsistencies herein, nor for damages resulting from performing the experiments.

Your use and/or purchase of this Manual implies your acceptance of this disclaimer.

YouTubes

Please note that several of these experiments are illustrated on YouTube. They can be accessed here: https://tinyurl.com/CrockerMicro

Bacteriology Experiment in the Home

Introduction: Our homes are surprisingly germy places. There are bacteria living on surfaces, in our water bottles, in our dishwasher, on our hands, on our feet, and on our phones. In some cases, there are many and, in others, only a few. In this introductory microbiology experiment, students will get a feel for what is around them. Be prepared to be surprised!

Purpose: To give students hands-on microbiology experience and to learn about bacteria around the home.

Materials:

- 10 Pre-poured nutrient or LB agar plates (**kept refrigerated** until used),
- tape,
- Sharpie,
- Q-tips,
- bottled or boiled cooled water,
- black plastic bag,
- items to be tested.

Method:

WITHOUT OPENING THE PLATES, observe that they have a lid and a bottom, where the agar gel, which includes nutrient broth (meat extract), is located. Currently, the inside of the plates should be sterile. Carefully observe whether there are any contaminants. Dispose of that plate. Turn the plates upside-down, with the nutrient agar at the top, and allow the plates to reach room temperature.

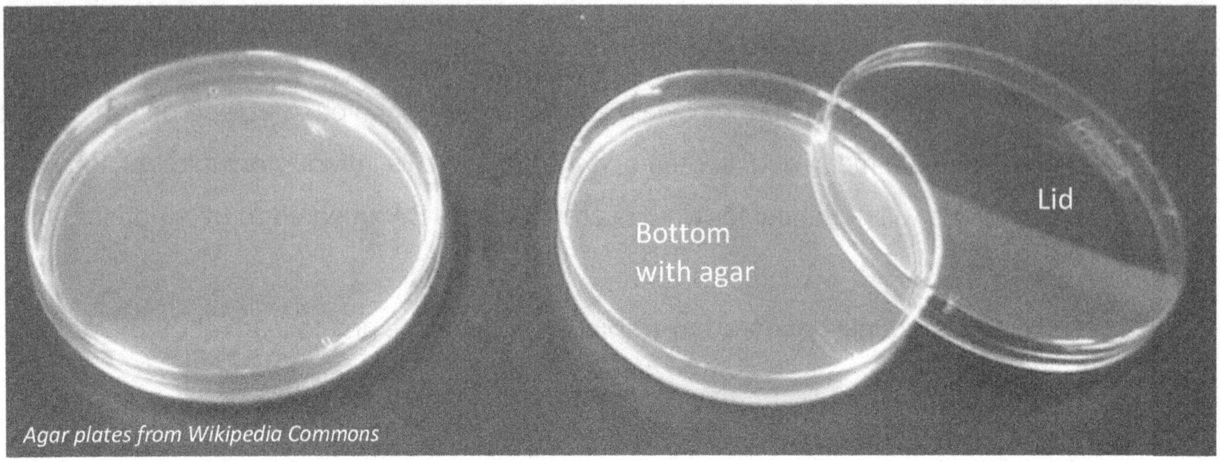

Agar plates from Wikipedia Commons

While the plates are sitting upside down, plan the places in your home that you would like to sample. Chose some that you think will be very germy and some that you think won't be. The floor, the sink, the counter, the toilet, the hairbrush, your hand before and after washing it, the doorknob, yogurt, meat, flower water, your mouth, your toe, your water bottle, your dog's mouth. Or use your imagination.

Write what will be in each plate on the BOTTOM of the plate with a Sharpie. DO NOT open the plates and do not turn them over.

Open the bottled or boiled cooled water. (The water is not sterile, but it's good enough for these purposes.) Wet a Q-tip in the water, swirl it over the first place you are testing, open the relevant plate, and gently swirl the tip of the Q-tip over the agar. Close the plate and put it upside down again. Repeat with a new Q-tip for each place you are testing. For your fingers or hair or something like that, there is no need for a Q-tip. Just press them gently onto the agar.

Now, tape around the edges of each plate to seal it. Do not obscure the top with tape. Because you might be growing pathogenic organisms, you will NOT take this tape off at any time. Ever.

Tape around the edges (shown in black) once the plate is inoculated

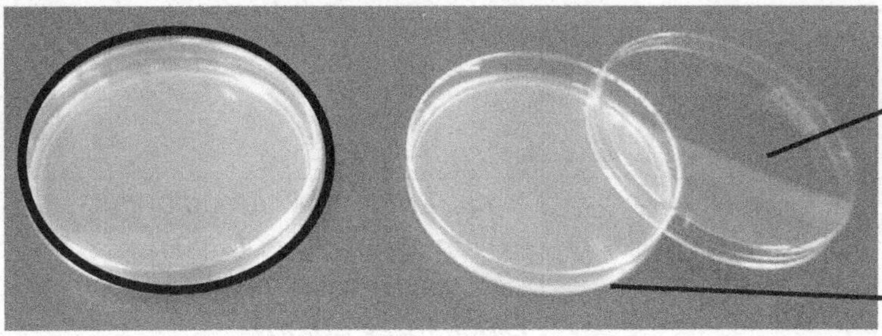

Do not write here

Write under here

Place the plates, upside down, in a black plastic bag (do not seal it) and put them somewhere warm indoors. Perhaps near a radiator, on top of the fridge, or in a sunny place. Do not put the black bag outside in the sun, since inside will reach temperatures that are too high for bacteria to survive. They should not exceed body temperature. Remember that bacteria like to grow upside-down; that prevents condensation from smearing the colony growth. Make sure they are not in reach of the dog, the cat, or the baby!

Leave the plates for two to four days. The length of time will depend on the temperature. Then, unwrap them and turn them over, but do not remove the tape, and DO NOT OPEN them. There should be no condensation at this point. They will smell bad, but are not dangerous provided that you do not open the lid. You will be able to see the bacterial colonies through the lid.

Observations:

Note what you see on each plate. Count the colonies (each comes from a single bacterium). Note their morphology, including color, size, shape, and whether they appear slimy. A different type of colony indicates a different species of microbe. Put this data into the table. Take photos if you like or draw each plate. Again, it may be tempting to take off the lid. Do not do so.

Place the plates back in the plastic bag, seal it, and dispose of it in the outside garbage can. Disinfect your workplace. Wash your hands thoroughly.

Results

Sample	Number of each type of colony	Description of colonies
For example: *Dog's ear*	*3* *5* *10* *1*	*Small smooth white colonies* *Yellow wrinkled colonies* *Large cream colonies* *Black fuzzy colony*

Analysis:

Why do you think some surfaces have more organisms than others?

Why did nothing grow on some plates? Does that mean the surface was sterile? What else could be happening? For example, would some organisms require more or different nutrients?

Do you see more or less growth from samples originating from dry surfaces? Would viruses grow here? How about parasites?

Do you suspect any of the colonies are fungal? Why?

How will the results modify your behavior, if they do?

Do any of the results raise questions in your mind?

Microbiology Experiments You Can Do at Home
Caroline Crocker, PhD

Mycology Experiment in the Home

Introduction: Fungi are all around us; we know them as mold, mildew, yeast, and mushrooms. Some grow best in the soil; some grow on different parts of plants. Some spoil our food; some help make our food. Some grow in our hair and some on our skin (Athlete's foot). Some are helpful (making bread) and some are not. Where do you think that fungi might thrive around your home?

Purpose: To give students hands-on mycology experience and to learn about fungi around the home.

Materials:

- 6 Pre-poured Sabouraud, Potato dextrose or Malt Extract agar plates (**kept refrigerated** until used),
- tape,
- Sharpie,
- Q-tips,
- black plastic bag,
- bottled or boiled cooled water
- items to be tested.

Method:

WITHOUT OPENING THE PLATES, observe that they have a lid and a bottom, where the agar gel, which contains nutrients for fungi and is slightly acidic, is located. Currently, the inside of the plates should be sterile. Carefully observe whether there are any contaminants. Dispose of that plate without opening it. Turn the plates upside-down with the agar at the top and allow them to reach room temperature.

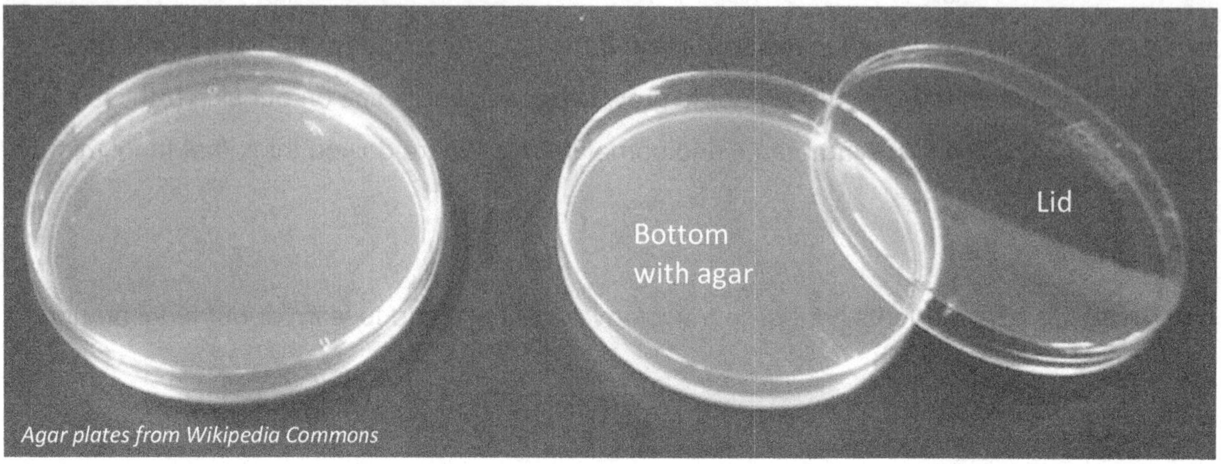

Agar plates from Wikipedia Commons

While the plates are sitting upside down, plan the places in and around your home that you think may have fungus growing. Examples could include soil from outside, a rose thorn or leaf, a root, an eyelash or hair, toe or fingernail cuttings, your dog's fur, moldy bread, water that flowers have been sitting in, beer or wine. Or use your imagination.

Write what will be in each plate on the BOTTOM of the plate with a Sharpie. For those plates that you will be using with solid samples, such as a nail or a hair, draw a line down the middle of the bottom of the plate and only use half for each sample. DO NOT open the plates and do not turn them over.

Tape around the edges (shown in black) once the plate is inoculated

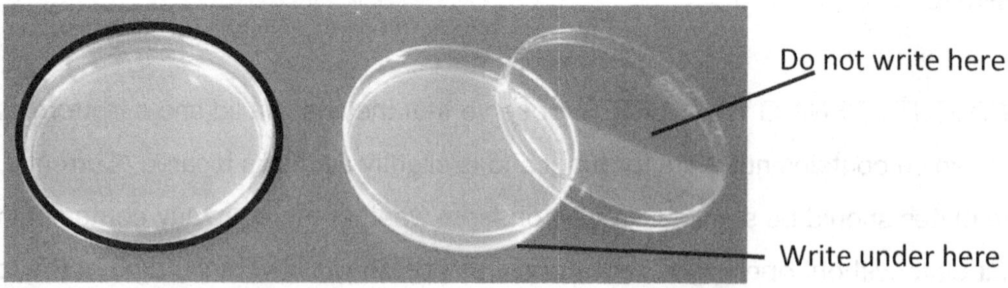

Open the bottled water. (The water is not sterile, but it's good enough for these purposes.) For soil, plant parts, hairs, and nails, just place them gently on the surface of the agar, being sure

that they are in contact with it by gently pressing with a Q-tip, so they do not fall off when the plate is turned upside down.

For the liquid samples, wet a Q-tip in them, open the relevant plate, and gently swirl the tip of the Q-tip over the agar. Close the plate and put it upside down again. Repeat with a new Q-tip for each new liquid that you are testing.

Now, tape around the edges of each plate to seal it. Do not obscure the top with tape. Because you might be growing pathogenic organisms and spored can become airborne, you will NOT take this tape off at any time. Ever.

Place the plates, upside down, in a black plastic bag (do not seal it) and put them somewhere warm indoors. Perhaps near a radiator, on top of the fridge, or in a sunny place. Do not put the black bag outside in the sun or anywhere it will exceed body temperature since inside will reach temperatures that are too high for anything to survive. Remember that microbes like to grow upside-down in the dark—well, this way the condensation will not obscure your results. Make sure the plates are not in reach of the dog, the cat, or the baby!

Leave the plates for three days. Then, unwrap them and turn them over, but do not remove the tape, and DO NOT OPEN them. There should be no condensation at this point. You will be able to see the fungal colonies through the lid. Take photos and then put the plates back to incubate for another four days. Take photos again.

Observations:

Note what you see on each plate. If possible, count the colonies (each comes from a single fungal organism). Note their morphology, including color, size, shape, and whether they appear slimy. A different type of colony indicates a different species of microbe and some will be bacteria. The fungi may appear fuzzy. Each 'spot' grows from one original organism. Put this data into the table below. Take photos if you like, or draw each plate. Again, it may be tempting to take off the lid. Do not do so.

After the second set of observations, place the plates back in the plastic bag, seal it, and dispose of it in the outside garbage can. Disinfect your workspace and wash your hands thoroughly.

Results after 3 days

Sample	Number of colonies	Description
Eg. Hair	*10* *3* *1*	*Smooth white tiny colonies* *Greenish fuzzy large colonies* *White fuzz with a black center*

Results after 7 days

Sample	Number of colonies	Description

Analysis:

Why do you think some samples yielded more fungi than others? Do you suspect some of the colonies were bacterial rather than fungal?

Why did nothing grow on some plates? Does that mean the surface was sterile? What else could be happening? For example, would some fungi require more or different nutrients?

Do you see more or less growth from samples originating from dry surfaces?

Did you see any additional growth or development after a week?

Did you see any mycelia? Any spores? Would they be visible?

How will the results modify your behavior, if they do?

Can you identify any of the fungi by comparing morphology with photos from online?

Do you think any of these fungi could be pathogenic? Why or why not?

Food Experiment in the Home

Introduction: Many foods are made with the help of microbes. Yogurt, whether advertised as live or not, has living cultures of various bacteria in it. Bread is risen by a fungus that digests the starch and makes CO_2: yeast. Cheese is made out of milk by the action of various microbes.

Very little of our food is sterile. Even pasteurized milk, while no longer containing active TB bacteria, still contains living organisms. That's why it goes sour eventually. Cream is the same. And, of course, we've all read about people being sickened by Salmonella that may contaminate eggs, *E. coli* in hamburger meat, and various organisms on fresh fruit and vegetables.

Purpose: To give students hands-on microbiology experience and to learn about organisms found in foods, whether as contaminants or as organisms that make the food what it is.

Materials:

- 4 Pre-poured Sabouraud, Potato dextrose or Malt Extract agar plates for fungi,
- 4 Pre-poured Nutrient Agar or LB plates for bacteria,
- an unopened bottle of water,
- tape,
- Sharpie, Q-tips,
- several clean cereal bowls (preferably washed in a dishwasher or rinsed with boiled water),
- black plastic bag,
- various foods.

<u>Keep the plates refrigerated</u> until they are used.

Method:

Figure out eight foods that you would like to test. These will be tested on BOTH types of agar. Examples could include hard or soft cheese, ham, milk, raw hamburger or chicken, egg, fruit, lettuce, spinach, yogurt, vinegar, cream, left-over takeout food, leftover rice or pasta, beer, etc. Or use your imagination. Any food is okay—maybe even try something from a jar or can.

WITHOUT OPENING THE PLATES, observe that they have a lid and a bottom, where the agar gel, specific to grow fungi or bacteria, is located. Currently, the inside of the plates should be sterile. Carefully observe whether there are any contaminants. Dispose of that plate. Turn the plates upside-down, with the agar at the top, and allow the eight plates to reach room temperature. Be sure that you know which are specific for fungi and which are for bacteria. Mark those for fungi "F" and those for bacteria "B", writing on the bottom outside of the bottom of the plate.

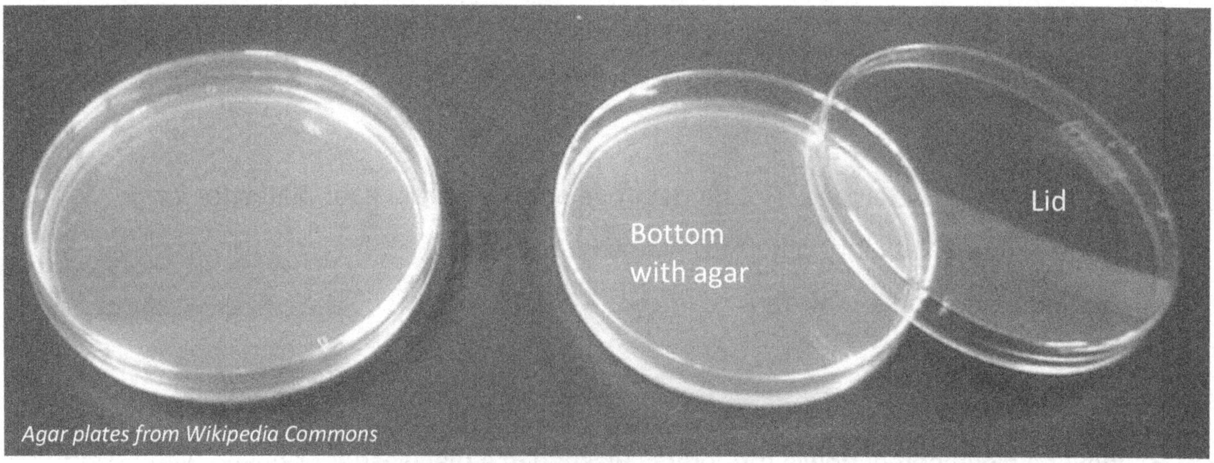

Agar plates from Wikipedia Commons

Divide each plate into two by drawing a line down the middle of the BOTTOM of the plate. Write what will be on each side of each plate on the BOTTOM of the plate with a Sharpie. Remember that you will be testing each of the eight types of food on both types of plates. See the examples below for nutrient agar (NA) and Sabouraud agar (SA). DO NOT open the plates and do not turn them over.

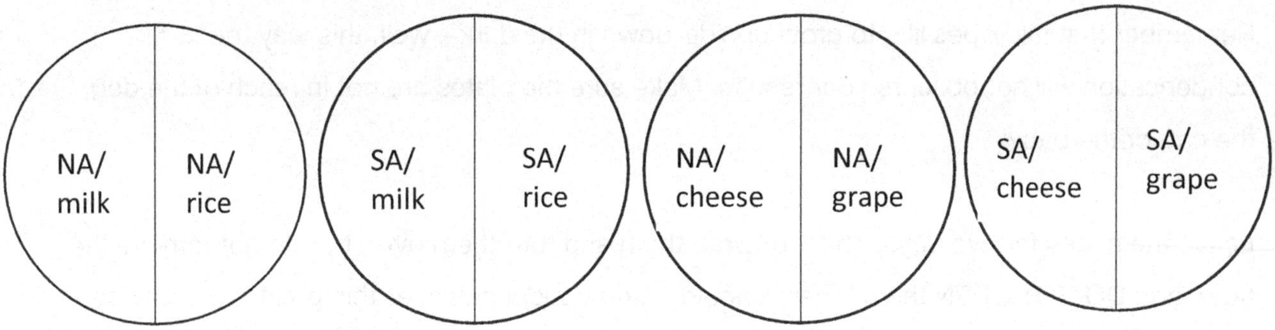

Open the bottled water. (The water is not sterile, but it's good enough for these purposes.) For solid foods, pour a small amount of the water into a clean bowl, add the food, and swirl it around with a Q-tip. Then gently swirl the Q-tip over half of the agar for bacteria in the right place. Dip the Q-tip in the water with the solid food again and swirl it over the correct half of the agar for fungi. Repeat for the other solid foods. Use a clean bowl for each type of food—they are not sterile but good enough for our purposes.

For the liquid foods, wet a Q-tip in them, open the relevant plate, and gently swirl the tip of the Q-tip over the proper side of the agar plate for bacteria. Use a clean Q-tip to wet it in the wet food again and swirl that over the surface of the agar plate for fungi. Close the plates and put them upside down again. Repeat with new Q-tips for each new liquid that you are testing.

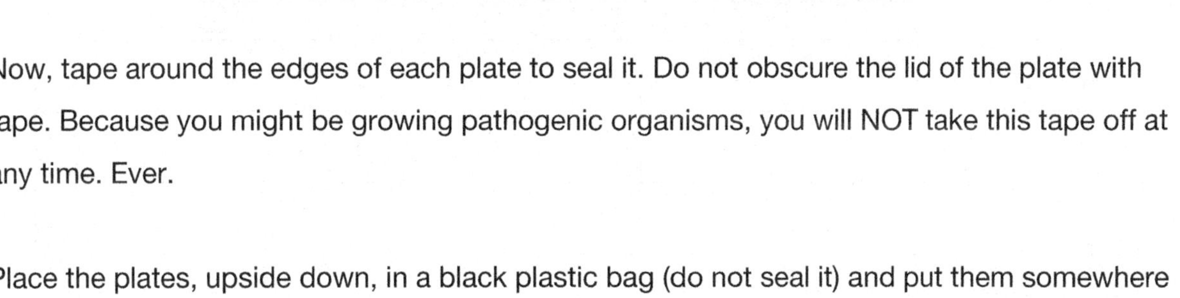

Now, tape around the edges of each plate to seal it. Do not obscure the lid of the plate with tape. Because you might be growing pathogenic organisms, you will NOT take this tape off at any time. Ever.

Place the plates, upside down, in a black plastic bag (do not seal it) and put them somewhere warm indoors. Perhaps near a radiator or in a sunny place. Do not put the black bag outside in the sun, since inside will reach temperatures that are too high for anything to survive. Remember that microbes like to grow upside-down in the dark—well, this way the condensation will not obscure your results. Make sure the plates are not in reach of the dog, the cat, or the baby!

Leave the plates for two days. Then, unwrap them and turn them over, but do not remove the tape, and DO NOT OPEN them. There should be no condensation at this point. You may be able to see the colonies through the lid. Take photos and then put the plates back to incubate for another two days. Take photos again.

Observations:

Note what you see on each plate. If possible, count the colonies (each comes from a single organism). Note their morphology, including color, size, shape, and whether they appear slimy or fuzzy. A different type of colony indicates a different species. But, remember that a colony is like a city of microbes, all with one parent. Although it grew from a single organism, it is now millions of them. Put your data into the table below. Take photos if you like, or draw each plate. Again, it may be tempting to take off the lid. Do not do so.

After the second set of observations, place the plates back in the plastic bag, seal it, and dispose of them in the outside garbage can. Disinfect your work surface and wash your hands thoroughly.

Results on ____________ Agar for Bacteria after 2 days

Food Type	Number of colonies	Description

Results on __________ Agar for Fungi after 2 days

Food Type	Number of colonies	Description

Results on __________ Agar for Bacteria after 4 days

Food Type	Number of colonies	Description

Results on ___________ Agar for Fungi after 4 days

Food Type	Number of colonies	Description

Analysis:

Why do you think some foods had more growth than others?

Why do you think nothing grew on some plates?

Why did some foods give growth on the plate for bacteria and not the one for fungi or vice versa? Think about the nutrients, temperature, salt, or sugar concentration needed for different organisms.

Do you see different types of growth from different foods?

Hypothesize about what species of organisms may be present in different foods.

If there was no growth, does that mean the food was sterile? Why or why not?

How will the results modify your behavior, if they do?

Do any of the results raise questions in your mind?

Microbiology Experiments You Can Do at Home
Caroline Crocker, PhD

Testing Antimicrobials in the Home

Introduction: Since microorganisms can be pathogenic, either by infecting us or by spoiling our food, it is important to know how best to control them. Disinfectants are used to control microbial growth on surfaces, and antiseptics are used for organisms found on the skin. We will be testing several of each.

Purpose: To give students hands-on microbiology experience and to learn about how effective various chemicals are at inhibiting microbial growth.

Materials:

- 4 Pre-poured MRS plates for Lactobacilli,
- unopened bottle of water,
- tape,
- Sharpie,
- tweezers,
- sheet of bonded paper or paper towel,
- hole punch,
- various disinfectants,
- black plastic bag,
- eight cups,
- spoon or rubber spatula,
- plain yogurt,
- tealight candle,
- metal cookie tin large enough to accommodate the plates.

Keep the plates refrigerated until the day that they are used. Warm them to room temperature before beginning the experiment.

Method:

Use the hole punch to make small circular pieces of paper. These will be dipped in your antimicrobial agents. Choose eight from the following list of disinfectants or antiseptics: diluted bleach (can try different concentrations), Clorox wipes, hand sanitizer, brandy or alcohol, an essential oil, hydrogen peroxide, antibiotic cream, crushed fresh garlic, macerated fresh oregano leaves, salt water, leftover antibiotics dissolved in water. Write down which antimicrobial agent will correspond to which number.

Place a tablespoon of plain yogurt in 0.5 cups of water. Stir it well.

WITHOUT OPENING THE PLATES, observe that they have a lid and a bottom, where the agar gel is located. Currently, the inside of the plates should be sterile. Carefully observe whether there are any contaminants. Dispose of that plate. Turn the plates upside-down, with the agar at the top, and allow the plates to reach room temperature.

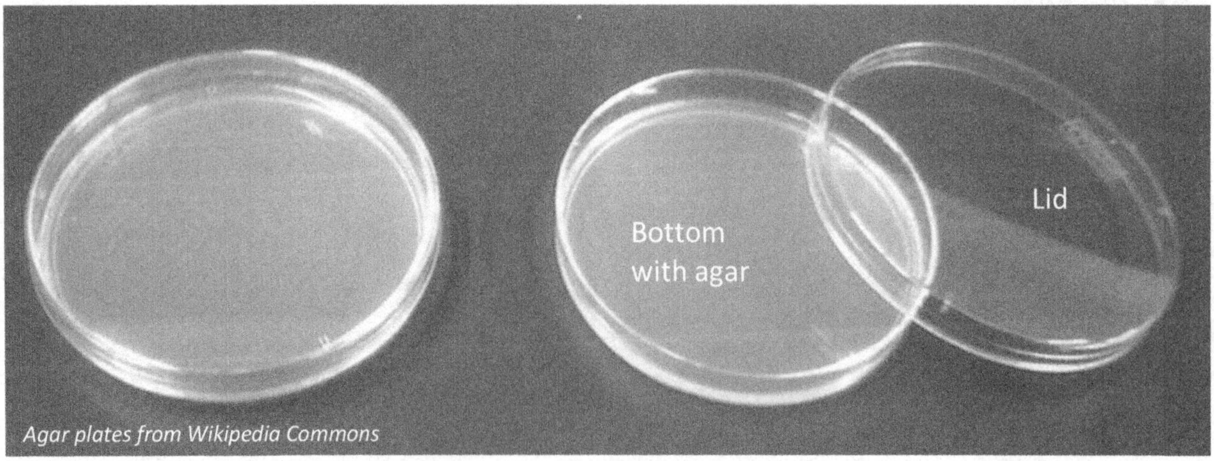

Agar plates from Wikipedia Commons

Take the agar-containing bottom off each plate and place it agar-up on a table. Put a few drops of yogurt water on each plate. Spread the water around using the back of a clean spoon or a rubber spatula, being sure to cover the entire plate as best you can. Shake off the excess fluid over the sink. Put the lids on the plates and leave for an hour for the water to be absorbed.

While you are waiting, prepare your antimicrobial agents. For those that are solid, macerate them and add a bit of water to them. For those that are liquid, place a few drops of each antimicrobial agent in a cup. You can also try different concentrations by adding a teaspoon to 3 tablespoons of water (1:10 dilution). Drop a disc of paper into each cup with antimicrobial agent. Leave for a few minutes. Be sure to write down which agent is in which cup!

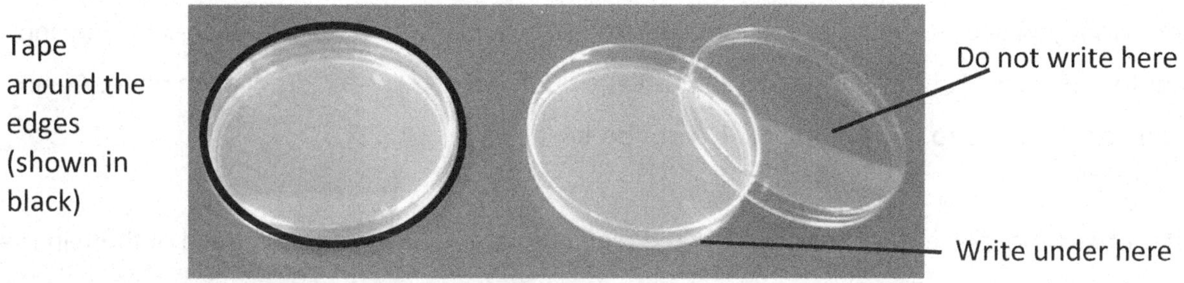

Turn the plates over so that you can write on the bottoms. Divide the plate into two halves by drawing a line on the bottom. Number each half of the bottom of the plates from 1 to 8. Then, using tweezers, pick up a disc and put it in the correct half of an agar plate. Rinse the tweezers off with water between each antimicrobial agent and repeat for the other plates. Replace the lids onto the plates immediately after placing the discs.

Since we do not want the discs to fall off, we will be incubating these plates right way up, with the lid on top and the agar on the bottom. Carefully tape around the edges of each plate to seal it, being sure not to move the discs inside. Do not obscure the top with tape. Because you might be growing pathogenic organisms, you will NOT take this tape off at any time. Ever.

Lactobacilli are capnophiles—they like CO_2. Place the plates, right way up, in a metal box (cake or cookie tin). Place a tealight candle inside the box and light it. DO NOT PUT THE CANDLE ON THE PLATES—they do melt. Replace the lid and tape around the edges. The candle will use up the oxygen and increase the concentration of CO_2 in the box. Put the box somewhere warm indoors, but be sure it does not exceed body temperature. Make sure the plates are not in reach of the dog, the cat, or the baby.

Leave the plates for three days. Then, unwrap them, but do not remove the tape, and DO NOT OPEN them. There should be no condensation at this point. You may be able to see the colonies through the lid. Hopefully, you will see clearing around some of the antimicrobial agents.

Observations:

Note what you see on each plate. If possible, measure how far/if the bacteria are not growing near to the discs. Put this data into a table. Take photos if you like, or draw each plate. Again, it may be tempting to take off the lid. Do not do so.

After observing, place the plates back in the plastic bag, seal them, and dispose of them in the outside garbage can. Disinfect your work area and wash your hands thoroughly.

<u>Results after Three Days of Exposure to Antimicrobials</u>

Antimicrobial agent/concentration	Zone of clearing

Analysis:

Why do you think some antimicrobials worked better than others?

Do you think that the result obtained with Lactobacillus would apply to other bacteria? To fungi? To viruses?

Did any of the results surprise you?

How will the results modify your behavior, if they do?

Do any of the results raise questions in your mind?

Microbiology Experiments You Can Do at Home
Caroline Crocker, PhD

Bacterial Plate Counting

Introduction: There are many times that scientists need to know how many microorganisms they have in suspension. How fast do microbes grow in this culture? How many bacteria are there in this environmental water sample? It is possible to count bacteria in a counting chamber using a microscope, but not at home! Also, that does not distinguish dead from viable organisms. The plate culture will not reveal the presence of dead bacteria, but it will enable the counting of live ones. Remember, each colony that you will see represents one live cell—one colony-forming unit (CFU).

Purpose: To give students hands-on microbiology experience and to learn about standard plate counting.

Materials:

- 6 Pre-poured MRS plates for bacteria,
- unopened bottle of water,
- tape,
- Sharpie,
- Teaspoon, ¼ tsp, and tablespoon measure,
- rubber spatula,
- metal cookie tin,
- six clean cups that have been rinsed out with boiling water,
- live yogurt,
- tealight candle.

Keep the plates refrigerated until the day that they are used.

Method:

Pour the bottle of water into a pan, add the teaspoon and the tablespoon measure, and boil them. Allow to cool. Remove the teaspoon and tablespoon using clean tongs.

The yogurt will be your source of bacteria—*Lactobacilli.* You will be determining how many microbes there are per ml of yogurt.

WITHOUT OPENING THE PLATES, observe that they have a lid and a bottom, where the agar gel is located. Currently, the inside of the plates should be sterile. Carefully observe whether there are any contaminants. Dispose of that plate. Turn the plates upside-down, with the agar at the top, and allow the plates to reach room temperature. Label the bottom of each plate #1, #2, #3, #4, #5, and #6.

Line the cups up and place 3 tablespoons of cooled boiled water in each. This is equivalent to 9 teaspoons. Add one teaspoon of yogurt and add it to the first cup. Stir with the teaspoon to mix up the solution. Take one teaspoon of the first dilution and add it to the second cup. Stir to mix and take one teaspoon and add to the third. Repeat until you have moved one teaspoon down the entire line of cups.

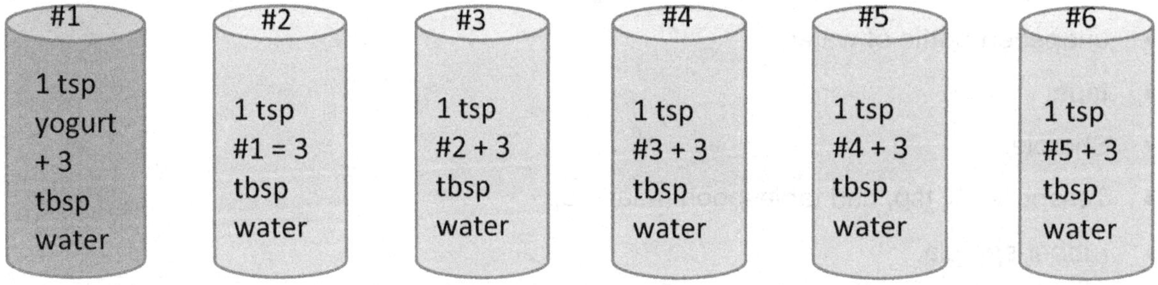

Now, # 1 has a 1 in 10 dilution of yogurt. #2 has a 1 in 100 dilution of yogurt. What dilution is in

#3? ___________

#4?___________

#5?___________

#6?___________

Take the agar-containing bottom off plate #6 and place it agar-up on a table. Put a ¼ teaspoon of the contents of cup #6 on the plate. Spread the water around using the spatula, being sure to cover the entire plate as best you can. Put the lid on the plate. Repeat with plate #5 and cup #5. Continue plating, working from most dilute to least.

Why do you think we plate in this order? Why would it not work to plate the other way?

Leave the plates with the lid side up for an hour for the water to be absorbed. Then tape around the edges as shown.

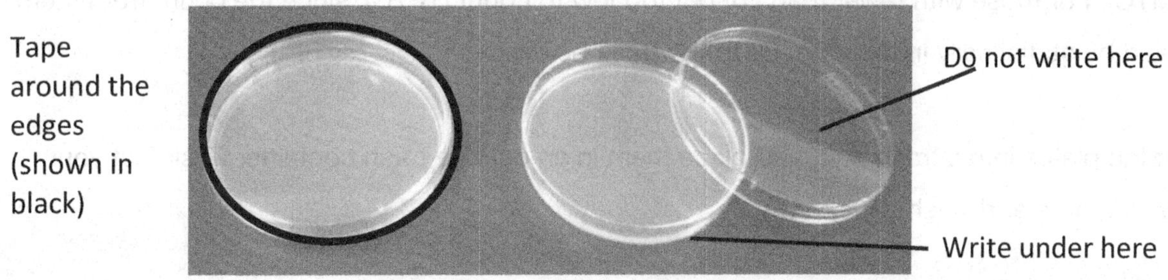

Do not obscure the top with tape. Although MRS agar is selective, making it unlikely that you will be growing pathogenic organisms, it is best to be safe. You will NOT take this tape off at any time. Ever.

Place the plates, upside down, in the cookie tin. Put the tealight candle in with them, BUT NOT ON THEM (they melt), light it, and replace the lid. Tape around the lid. The candle will burn the oxygen and make the atmosphere CO_2-enriched. Put the box somewhere warm, but not above body temperature. Do not put the box outside in the sun, since inside will reach temperatures that are too high for anything to survive. Make sure the plates are not in reach of the dog, the cat, or the baby.

Leave the plates for three days. Then, unwrap them, but do not remove the tape, and DO NOT OPEN them. There should be no condensation at this point. You may be able to see the colonies through the lid.

Observations:

Note what you see on each plate. Count the bacterial colonies. Some may be small, so be sure to have good light. Take photos if you like, or draw each plate. Again, it may be tempting to take off the lid. Do not do so.

Only count the CFU on plates with between 20-30 and 300 colonies. It may be helpful to do this by suspending the plate over a light, looking through the bottom of the plate, and marking those you count with the Sharpie. For those with more than 300, put too many to count (TMTC). For those with fewer than 20, put too few to count (TFTC), since the count from them would be statistically inaccurate. Fill in the table below.

Put the plates into a trash bag, and place them in an outside trash container. Disinfect your working area and wash your hands.

Results

Cup Number	Dilution	Number of CFU
1	1/10	
2	1/100	
3		
4		
5		
6		

Analysis:

In science, we do not measure in teaspoons, but in ml. Fortunately, the conversion is possible. There are 5 ml in a teaspoon and you used ¼ tsp.

To convert from one set of units to another, you essentially need to multiply by one. You are starting with CFU/plate. You want to know CFU/ml yogurt. Here are the facts you work with:

- 1 plate is inoculated with ¼ tsp, so multiplying by 1plate/1/4 tsp is like multiplying by 1
- 1 tsp contains 5 ml, so multiplying by 1 tsp/5 ml is like multiplying by 1
- Cup 5 had a concentration of yogurt of 1/100,000. So there would have been $1/10^5$ ml of yogurt in 1 ml of the dilution. So one could multiply by 1 ml/$1/10^5$ ml yogurt and it would be like multiplying by 1.

I had 24 colonies on plate #5. To calculate the CFU/ml in yogurt, I did the following:
24 CFU/plate x 1 plate/¼ tsp x 1 tsp/5 ml x 1 ml diluted/1/100,000 ml yogurt = 1,920,000 CFU/ml yogurt or 1.9×10^6 CFU/ml yogurt in scientific notation. (Remember that dividing by a fraction is the same as multiplying by the inverse of the fraction.) Note that you can cancel out the units and you end up with CFU/ml yogurt. That is how you know that your multipliers are the right way up!

How many CFU were there in a tsp of your diluted yogurt? ______________________________

How many CFU were there in a ml of the diluted yogurt? ______________________________

How many CFU/ml were in your undiluted yogurt? ______________________________________

What is that in scientific notation? ______________________________

How do you know which way up to put your conversion factors? ______________

____

Why do we not report this as bacteria/ml? ______________________________

____

It may be that some instructors and homeschool teachers would like their students to complete the microbiology experience by writing a lab report. Herein follows the report format that I generally advise students to use.

Lab Report Format

Abstract

An abstract is a paragraph at the beginning of a research manuscript that summarizes the contents of the entire report. It begins with a one or two-sentence statement about the aim of the investigation. The following sentence or two are about the methods employed. Next, the results are briefly summarized and, finally, the significance of the results is stated in a tentative fashion (as a conclusion).

Introduction

In this section, the reader is introduced to the experiment by being acquainted with previous research results in the particular area of interest. Out of this, the authors show the problem that they are attempting to solve. This is where the background would be explained. The last paragraph of the introduction is generally a statement of the specific purpose of the experiments or the hypothesis.

Methods

In published research, this section describes the methods in detail so that others can repeat the work. However, if the methods have been described elsewhere, it is common practice to refer the reader to that document with a statement such as: proteins were purified as described in Doe *et al*. Similarly, it is acceptable, for your lab reports, to briefly describe the methodology and refer the reader to this booklet for details. DO NOT copy word-for-word. Be sure to write in the past tense, reporting what you did, not as if you are giving others a list of instructions.

Results

The results section of a paper documents what was observed in the form of text, tables, photographs, drawings, and graphs. This is where you would say what kinds of colonies were found on each plate and how many there were. In addition, you may wish to include anything of interest that you observed. It is very tempting for scientists to falsify data (especially if it does not seem right), however, we are well aware that experiments do not work every time and honesty is an essential quality in the production of good scientific work! Be careful not to interpret your results; that is for the discussion.

Discussion

In this section of the paper, you may interpret your results and suggest what they mean. Did any of the results surprise you? Why do you think one plate had more colonies than another? Do you think that this method would show ALL possible types of bacteria? If you had problems, this is where you should speculate on why the experiment did not work and how it could have been improved.

References

It is essential when attempting publication of one's results that every statement is backed up with a reference. However, since you are unlikely to publish your lab reports, here you may just list the references you consulted to write the introduction and figure out what your results may mean.

Made in the USA
Monee, IL
08 November 2026

67994713R00026